"The next best thing to knowing something is knowing where to find it" -Samuel Johnson

"

This Planner Belongs To :

....

Yearly Assignment Planner

Top Three Goals:

1

2

3

Month

Month

Month

Month

Notes:

Yearly Assignment Planner

Month

Month

Month

Month

Month

Month

Notes:

Yearly Assignment Planner

Month

Month

Achievements

....
....
....
....
....
....
....
....
....
....
....
....
....
....
....
....
....
....
....

Pending Assignments

....
....
....
....
....
....
....
....
....
....
....
....
....
....
....
....
....
....
....

Notes :

Monthly Assignment Planner

Achievements

| |
| |
| |
| |
| |
| |

Pending Assignments

| |
| |
| |
| |
| |
| |

Chapters

| |
| |
| |
| |
| |
| |

Slow Learners

Notes:

Monthly Assignment Planner

Achievements

Pending Assignments

Chapters

Slow Learners

Notes:

Monthly Assignment Planner

Achievements

| |
| |
| |
| |
| |
| |

Pending Assignments

| |
| |
| |
| |
| |
| |

Chapters

| |
| |
| |
| |
| |
| |

Slow Learners

Notes:

2019 calendar

January 2019

SUN	MON	TUE	WED	THU	FRI	SAT
		1	2	3	4	5
6	7	8	9	10	11	12
13	14	15	16	17	18	19
20	21	22	23	24	25	26
27	28	29	30	31		

February 2019

SUN	MON	TUE	WED	THU	FRI	SAT
					1	2
3	4	5	6	7	8	9
10	11	12	13	14	15	16
17	18	19	20	21	22	23
24	25	26	27	28		

March 2019

SUN	MON	TUE	WED	THU	FRI	SAT
					1	2
3	4	5	6	7	8	9
10	11	12	13	14	15	16
17	18	19	20	21	22	23
24	25	26	27	28	29	30
31						

April 2019

SUN	MON	TUE	WED	THU	FRI	SAT
	1	2	3	4	5	6
7	8	9	10	11	12	13
14	15	16	17	18	19	20
21	22	23	24	25	26	27
28	29	30				

May 2019

SUN	MON	TUE	WED	THU	FRI	SAT
			1	2	3	4
5	6	7	8	9	10	11
12	13	14	15	16	17	18
19	20	21	22	23	24	25
26	27	28	29	30	31	

June 2019

SUN	MON	TUE	WED	THU	FRI	SAT
						1
2	3	4	5	6	7	8
9	10	11	12	13	14	15
16	17	18	19	20	21	22
23	24	25	26	27	28	29
30						

July 2019

SUN	MON	TUE	WED	THU	FRI	SAT
	1	2	3	4	5	6
7	8	9	10	11	12	13
14	15	16	17	18	19	20
21	22	23	24	25	26	27
28	29	30	31			

August 2019

SUN	MON	TUE	WED	THU	FRI	SAT
				1	2	3
4	5	6	7	8	9	10
11	12	13	14	15	16	17
18	19	20	21	22	23	24
25	26	27	28	29	30	31

September 2019

SUN	MON	TUE	WED	THU	FRI	SAT
1	2	3	4	5	6	7
8	9	10	11	12	13	14
15	16	17	18	19	20	21
22	23	24	25	26	27	28
29	30					

October 2019

SUN	MON	TUE	WED	THU	FRI	SAT
		1	2	3	4	5
6	7	8	9	10	11	12
13	14	15	16	17	18	19
20	21	22	23	24	25	26
27	28	29	30	31		

November 2019

SUN	MON	TUE	WED	THU	FRI	SAT
					1	2
3	4	5	6	7	8	9
10	11	12	13	14	15	16
17	18	19	20	21	22	23
24	25	26	27	28	29	30

December 2019

SUN	MON	TUE	WED	THU	FRI	SAT
1	2	3	4	5	6	7
8	9	10	11	12	13	14
15	16	17	18	19	20	21
22	23	24	25	26	27	28
29	30	31				

Notes

2019 Holidays

Jan 01	New Year's Day
Jan 21	Martin Luther King Day
Feb 05	Chinese New Year
Feb 12	Lincoln's Birthday
Feb 14	Valentine's Day
Feb 18	President's Day
Mar 06	Ash Wednesday
Mar 10	Daylight Saving (begin)
Mar 17	St. Patrick's Day
Mar 20	Vernal equinox (GMT)
Apr 01	April Fool's Day
Apr 20	Passover
Apr 21	Easter
Apr 24	Admin Assistants Day
May 06	Ramadan begins
May 12	Mother's Day
May 27	Memorial Day
Jun 09	Pentecost
Jun 14	Flag Day
Jun 16	Father's Day
Jun 21	June Solstice (GMT)
Jul 04	Independence Day
Sep 02	Labor Day
Sep 23	Autumnal equinox (GMT)
Sep 30	Rosh Hashanah
Oct 14	Columbus Day
Oct 31	Halloween
Nov 03	Daylight Saving (end)
Nov 11	Veterans Day
Nov 28	Thanksgiving
Dec 22	Hanukkah begins
Dec 22	December Solstice (GMT)
Dec 25	Christmas Day
Dec 26	Kwanzaa begins
Dec 31	New Year's Eve

2020 calendar

January 2020

SUN	MON	TUE	WED	THU	FRI	SAT
			1	2	3	4
5	6	7	8	9	10	11
12	13	14	15	16	17	18
19	20	21	22	23	24	25
26	27	28	29	30	31	

February 2020

SUN	MON	TUE	WED	THU	FRI	SAT
						1
2	3	4	5	6	7	8
9	10	11	12	13	14	15
16	17	18	19	20	21	22
23	24	25	26	27	28	29

March 2020

SUN	MON	TUE	WED	THU	FRI	SAT
1	2	3	4	5	6	7
8	9	10	11	12	13	14
15	16	17	18	19	20	21
22	23	24	25	26	27	28
29	30	31				

April 2020

SUN	MON	TUE	WED	THU	FRI	SAT
			1	2	3	4
5	6	7	8	9	10	11
12	13	14	15	16	17	18
19	20	21	22	23	24	25
26	27	28	29	30		

May 2020

SUN	MON	TUE	WED	THU	FRI	SAT
					1	2
3	4	5	6	7	8	9
10	11	12	13	14	15	16
17	18	19	20	21	22	23
24	25	26	27	28	29	30
31						

June 2020

SUN	MON	TUE	WED	THU	FRI	SAT
	1	2	3	4	5	6
7	8	9	10	11	12	13
14	15	16	17	18	19	20
21	22	23	24	25	26	27
28	29	30				

July 2020

SUN	MON	TUE	WED	THU	FRI	SAT
			1	2	3	4
5	6	7	8	9	10	11
12	13	14	15	16	17	18
19	20	21	22	23	24	25
26	27	28	29	30	31	

August 2020

SUN	MON	TUE	WED	THU	FRI	SAT
						1
2	3	4	5	6	7	8
9	10	11	12	13	14	15
16	17	18	19	20	21	22
23	24	25	26	27	28	29
30	31					

September 2020

SUN	MON	TUE	WED	THU	FRI	SAT
		1	2	3	4	5
6	7	8	9	10	11	12
13	14	15	16	17	18	19
20	21	22	23	24	25	26
27	28	29	30			

October 2020

SUN	MON	TUE	WED	THU	FRI	SAT
				1	2	3
4	5	6	7	8	9	10
11	12	13	14	15	16	17
18	19	20	21	22	23	24
25	26	27	28	29	30	31

November 2020

SUN	MON	TUE	WED	THU	FRI	SAT
1	2	3	4	5	6	7
8	9	10	11	12	13	14
15	16	17	18	19	20	21
22	23	24	25	26	27	28
29	30					

December 2020

SUN	MON	TUE	WED	THU	FRI	SAT
		1	2	3	4	5
6	7	8	9	10	11	12
13	14	15	16	17	18	19
20	21	22	23	24	25	26
27	28	29	30	31		

Notes

2020 Holidays

Jan 01	New Year's Day
Jan 20	Martin Luther King Day
Jan 25	Chinese New Year
Feb 12	Lincoln's Birthday
Feb 14	Valentine's Day
Feb 17	President's Day
Feb 26	Ash Wednesday
Mar 08	Daylight Saving (begin)
Mar 17	St. Patrick's Day
Mar 20	Vernal equinox (GMT)
Apr 01	April Fool's Day
Apr 09	Passover
Apr 12	Easter
Apr 22	Admin Assistants Day
Apr 24	Ramadan begins
May 10	Mother's Day
May 25	Memorial Day
May 31	Pentecost
Jun 14	Flag Day
Jun 20	June Solstice (GMT)
Jun 21	Father's Day
Jul 04	Independence Day
Sep 07	Labor Day
Sep 19	Rosh Hashanah
Sep 22	Autumnal equinox (GMT)
Oct 12	Columbus Day
Oct 31	Halloween
Nov 01	Daylight Saving (end)
Nov 11	Veterans Day
Nov 26	Thanksgiving
Dec 10	Hanukkah begins
Dec 21	December Solstice (GMT)
Dec 25	Christmas Day
Dec 26	Kwanzaa begins
Dec 31	New Year's Eve

Important DATES

AUGUST

_____
_____
_____
_____
_____
_____
_____
_____
_____
_____
_____

SEPTEMBER

Labor Day:02 _____
_____
_____
_____
_____
_____
_____
_____
_____
_____
_____

OCTOBER

Columbus Day:14 _____
Halloween:31 _____
_____
_____
_____
_____
_____
_____
_____
_____

FEBRUARY

Valentine's Day:14 _____
Presidents' Day:17 _____
_____
_____
_____
_____
_____
_____
_____
_____
_____

MARCH

St.Patrick's Day:17 _____
First Day of Spring:20 ___
_____
_____
_____
_____
_____
_____
_____
_____
_____

APRIL

Easter Day:12 _____
_____
_____
_____
_____
_____
_____
_____
_____
_____
_____

NOVEMBER

Veterans Day:11

Thanksgiving Day:28

DECEMBER

First Day of Winter:22

Christmas Eve:24

Christmas:25

New Year's Eve:31

JANUARY

New Year's Day:01

M L King Day:20

MAY

Cinco de Mayo:05

National TeacherDay16

Mother's Day:10

Memorial Day:25

JUNE

Father's Day:21

First Day of Summer:21

JULY

Independence Day:04

Student

	Student Name	Parent/Guardian	Phone: Home/Work/Cell
1			
2			
3			
4			
5			
6			
7			
8			
9			
10			
11			
12			
13			
14			
15			
16			
17			
18			
19			
20			
21			
22			
23			
24			
25			
26			
27			
28			
29			
30			
31			
32			
33			
34			
35			
36			

Roster

Address	E-mail/Fax	Medical Information	Bus Assignment

ATTENDANCE OR GRADES																							
NAME																							

ATTENDANCE OR GRADES																														
NAME																															

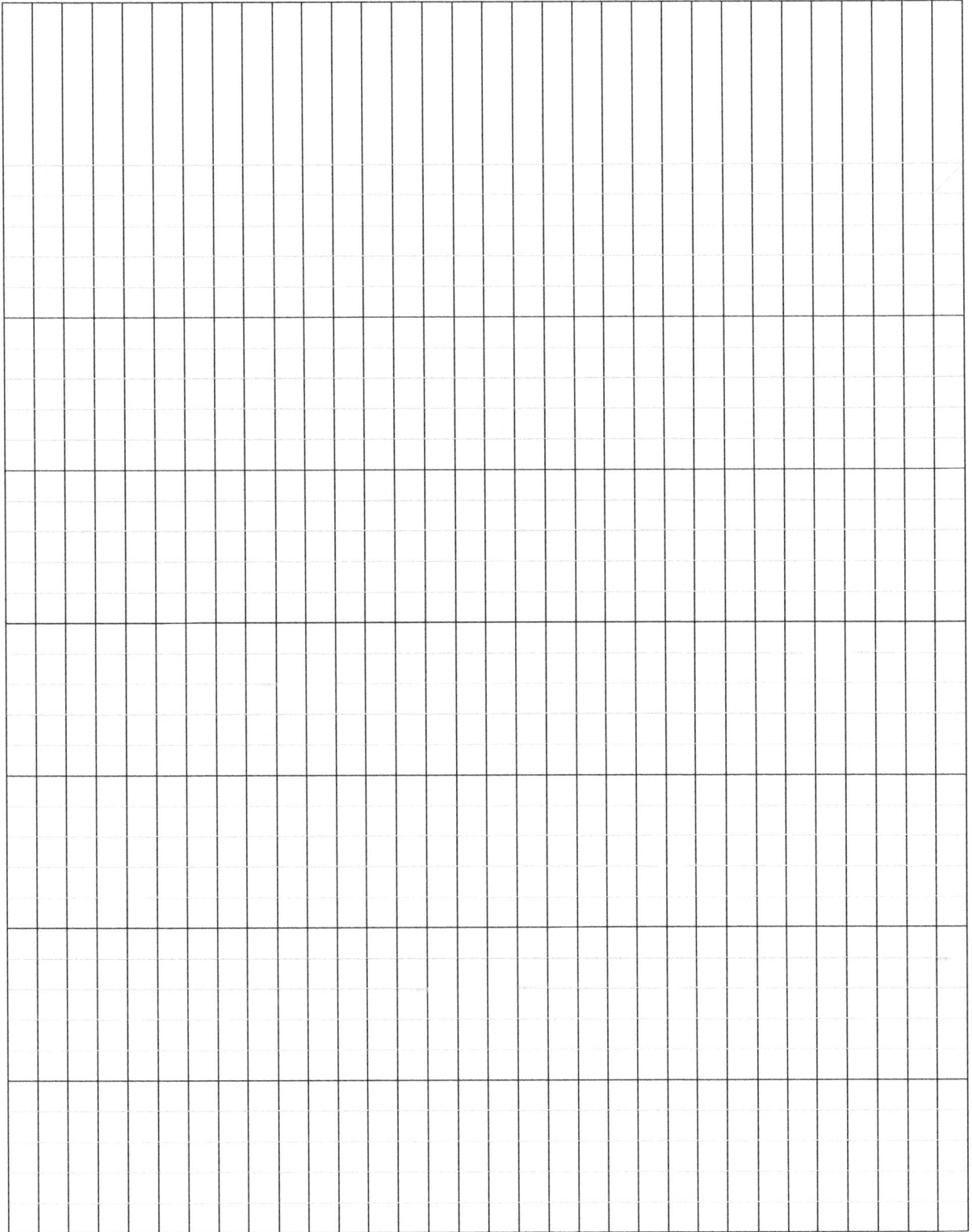

CALL HOME *Log*

TO CALL	REASON	DATE	RESOLVED?

CALL HOME *Log*

TO CALL	REASON	DATE	RESOLVED?

Rules

Rewards

Consequences

Student BIRTHDAYS

September	October	November
December	**January**	**February**
March	**April**	**May**
June	**July**	**August**

Seating Charts

Class_____ Period_____

Class_____ Period_____

SUBSTITUTE *Teacher* INFORMATION

Schedule

People to contact for help

Teacher:	
Student Helpers:	
Principal:	
Secretary:	
Custodian:	
Nurse:	
Other:	

Classroom procedures

Attendance:
Lunch:
Bathroom:
Recess:
Dismissal:
Other:

Attached to this information sheet you will find...

❏ Lesson Plans   ❏ School Rules   Other

❏ Class List   ❏ Bus List   ❏ _____

❏ Seating Chart   ❏ Map of school   ❏ _____

Class List

Period/Subject _____   Period/Subject _____

July 2019

SUN	MON	TUE	WED
30	1	2	3
7	8	9	10
14	15	16	17
21	22	23	24
28	29	30	31
4	5	6	7

July 2019

THU	FRI	SAT
4 *Independence Day*	5	6
11	12	13
18	19	20
25	26	27
1	2	3
8	9	10

June 2019

SUN	MON	TUE	WED	THU	FRI	SAT
						1
2	3	4	5	6	7	8
9	10	11	12	13	14	15
16	17	18	19	20	21	22
23	24	25	26	27	28	29
30						

notes

_____
_____
_____
_____
_____
_____
_____
_____
_____
_____
_____
_____
_____

August 2019

SUN	MON	TUE	WED	THU	FRI	SAT
				1	2	3
4	5	6	7	8	9	10
11	12	13	14	15	16	17
18	19	20	21	22	23	24
25	26	27	28	29	30	31

WEEK OF :.................

SUBJECT...........................	MONDAY Date............................	TUESDAY Date............................

WEDNESDAY Date............	THURSDAY Date...............	FRIDAY Date...............

WEEK OF :……………..

SUBJECT………………………	MONDAY Date………………………	TUESDAY Date………………………

WEDNESDAY Date............	THURSDAY Date...............	FRIDAY Date...............

WEEK OF :..................

SUBJECT...........................	MONDAY Date............................	TUESDAY Date............................
	_____	_____
	_____	_____
	_____	_____
	_____	_____

WEDNESDAY Date............	THURSDAY Date...............	FRIDAY Date...............

WEEK OF :………………..

SUBJECT……………………….	MONDAY Date…………………………	TUESDAY Date……………………………

WEDNESDAY Date............	THURSDAY Date...............	FRIDAY Date...............

WEEK OF :..................

SUBJECT.........................	MONDAY Date.........................	TUESDAY Date.........................

WEDNESDAY Date............	THURSDAY Date...............	FRIDAY Date...............
_____	_____	_____
_____	_____	_____
_____	_____	_____
_____	_____	_____
_____	_____	_____
_____	_____	_____
_____	_____	_____
_____	_____	_____
_____	_____	_____
_____	_____	_____
_____	_____	_____
_____	_____	_____
_____	_____	_____
_____	_____	_____
_____	_____	_____
_____	_____	_____
_____	_____	_____
_____	_____	_____
_____	_____	_____
_____	_____	_____
_____	_____	_____
_____	_____	_____
_____	_____	_____
_____	_____	_____
_____	_____	_____
_____	_____	_____
_____	_____	_____
_____	_____	_____

August 2019

SUN	MON	TUE	WED
28	29	30	31
4	5	6	7
11	12	13	14
18	19	20	21
25	26	27	28
1	2	3	4

August 2019

THU	FRI	SAT
1	2	3
8	9	10
15	16	17
22	23	24
29	30	31
5	6	7

July 2019

SUN	MON	TUE	WED	THU	FRI	SAT
	1	2	3	4	5	6
7	8	9	10	11	12	13
14	15	16	17	18	19	20
21	22	23	24	25	26	27
28	29	30	31			

notes

_____
_____
_____
_____
_____
_____
_____
_____
_____
_____
_____
_____
_____
_____

September 2019

SUN	MON	TUE	WED	THU	FRI	SAT
1	2	3	4	5	6	7
8	9	10	11	12	13	14
15	16	17	18	19	20	21
22	23	24	25	26	27	28
29	30					

WEEK OF :.................

SUBJECT...........................	MONDAY Date............................	TUESDAY Date............................

WEDNESDAY Date............	THURSDAY Date...............	FRIDAY Date...............
_____	_____	_____
_____	_____	_____
_____	_____	_____
_____	_____	_____
_____	_____	_____
_____	_____	_____
_____	_____	_____
_____	_____	_____
_____	_____	_____
_____	_____	_____
_____	_____	_____
_____	_____	_____
_____	_____	_____
_____	_____	_____
_____	_____	_____
_____	_____	_____
_____	_____	_____
_____	_____	_____
_____	_____	_____
_____	_____	_____
_____	_____	_____
_____	_____	_____
_____	_____	_____
_____	_____	_____
_____	_____	_____

WEEK OF :...................

SUBJECT...........................	MONDAY Date...........................	TUESDAY Date...........................

WEDNESDAY Date............	THURSDAY Date...............	FRIDAY Date...............

WEEK OF :..................

SUBJECT............................	MONDAY Date............................	TUESDAY Date............................

WEDNESDAY Date............	THURSDAY Date...............	FRIDAY Date...............

WEEK OF :……………..

SUBJECT……………………..	MONDAY Date…………………………….	TUESDAY Date…………………………….

WEDNESDAY Date...........	THURSDAY Date...............	FRIDAY Date...............

September 2019

SUN	MON	TUE	WED
1	2 *Labor Day*	3	4
8	9	10	11
15	16	17	18
22	23	24	25
29	30	1	2
6	7	8	9

September 2019

THU	FRI	SAT
5	6	7
12	13	14
19	20	21
26	27	28
3	4	5
10	11	12

August 2019

SUN	MON	TUE	WED	THU	FRI	SAT
				1	2	3
4	5	6	7	8	9	10
11	12	13	14	15	16	17
18	19	20	21	22	23	24
25	26	27	28	29	30	31

notes

_____

_____

_____

_____

_____

_____

_____

_____

_____

_____

_____

_____

October 2019

SUN	MON	TUE	WED	THU	FRI	SAT
		1	2	3	4	5
6	7	8	9	10	11	12
13	14	15	16	17	18	19
20	21	22	23	24	25	26
27	28	29	30	31		

WEEK OF :...................

SUBJECT..........................	MONDAY Date...........................	TUESDAY Date...........................

WEDNESDAY Date............	THURSDAY Date...............	FRIDAY Date................

WEEK OF :....................

SUBJECT...........................	MONDAY Date.............................	TUESDAY Date.............................

WEDNESDAY Date............	THURSDAY Date...............	FRIDAY Date...............

WEEK OF :..................

SUBJECT........................	MONDAY Date............................	TUESDAY Date............................

WEDNESDAY Date............	THURSDAY Date................	FRIDAY Date................

WEEK OF :..................

SUBJECT...........................	MONDAY Date.............................	TUESDAY Date.............................

WEDNESDAY Date............	THURSDAY Date...............	FRIDAY Date...............

WEEK OF :...................

SUBJECT...........................	MONDAY Date.............................	TUESDAY Date.............................

WEDNESDAY Date............	THURSDAY Date...............	FRIDAY Date...............

October 2019

SUN	MON	TUE	WED
29	30	1	2
6	7	8	9
13	14 *Columbus Day*	15	16
20	21	22	23
27	28	29	30
3	4	5	6

October 2019

THU	FRI	SAT
3	4	5
10	11	12
17	18	19
24	25	26
31 *Halloween*	1	2
7	8	9

September 2019

SUN	MON	TUE	WED	THU	FRI	SAT
1	2	3	4	5	6	7
8	9	10	11	12	13	14
15	16	17	18	19	20	21
22	23	24	25	26	27	28
29	30					

notes

_____

_____

_____

_____

_____

_____

_____

_____

_____

_____

_____

_____

November 2019

SUN	MON	TUE	WED	THU	FRI	SAT
					1	2
3	4	5	6	7	8	9
10	11	12	13	14	15	16
17	18	19	20	21	22	23
24	25	26	27	28	29	30

WEEK OF :………………..

SUBJECT………………………..	MONDAY Date………………………….	TUESDAY Date………………………….

WEDNESDAY Date...........	THURSDAY Date...............	FRIDAY Date...............

WEEK OF :.................

SUBJECT............................	MONDAY Date..............................	TUESDAY Date..............................

WEDNESDAY Date...........	THURSDAY Date...............	FRIDAY Date...............
_____	_____	_____
_____	_____	_____
_____	_____	_____
_____	_____	_____
_____	_____	_____
_____	_____	_____
_____	_____	_____
_____	_____	_____
_____	_____	_____
_____	_____	_____
_____	_____	_____
_____	_____	_____
_____	_____	_____
_____	_____	_____
_____	_____	_____
_____	_____	_____
_____	_____	_____
_____	_____	_____
_____	_____	_____
_____	_____	_____
_____	_____	_____
_____	_____	_____
_____	_____	_____
_____	_____	_____
_____	_____	_____

WEEK OF :.................

SUBJECT...........................	MONDAY Date...........................	TUESDAY Date...........................

WEDNESDAY Date...........	THURSDAY Date...............	FRIDAY Date...............

WEEK OF :..................

SUBJECT..........................	MONDAY Date.............................	TUESDAY Date.............................

WEDNESDAY Date............	THURSDAY Date................	FRIDAY Date................

November 2019

SUN	MON	TUE	WED
27	28	29	30
3	4	5	6
10	11 *Veterans Day*	12	13
17	18	19	20
24	25	26	27
1	2	3	4

November 2019

THU	FRI	SAT
31	1	2
7	8	9
14	15	16
21	22	23
28 *Thanksgiving Day*	29 *Black Friday*	30
5	6	7

October 2019

SUN	MON	TUE	WED	THU	FRI	SAT
		1	2	3	4	5
6	7	8	9	10	11	12
13	14	15	16	17	18	19
20	21	22	23	24	25	26
27	28	29	30	31		

notes

_____
_____
_____
_____
_____
_____
_____
_____
_____
_____
_____
_____

December 2019

SUN	MON	TUE	WED	THU	FRI	SAT
1	2	3	4	5	6	7
8	9	10	11	12	13	14
15	16	17	18	19	20	21
22	23	24	25	26	27	28
29	30	31				

WEEK OF :.................

SUBJECT.........................	MONDAY Date............................	TUESDAY Date............................

WEDNESDAY Date...........	THURSDAY Date...............	FRIDAY Date...............

WEEK OF :………………..

SUBJECT……………………….	MONDAY Date…………………………	TUESDAY Date…………………………

WEDNESDAY Date...........	THURSDAY Date...............	FRIDAY Date...............

WEEK OF :..................

SUBJECT.........................	MONDAY Date.............................	TUESDAY Date.............................

WEDNESDAY Date...........	THURSDAY Date...............	FRIDAY Date...............

WEEK OF :..................

SUBJECT...........................	MONDAY Date............................	TUESDAY Date............................

WEDNESDAY Date...........	THURSDAY Date...............	FRIDAY Date................

December 2019

SUN	MON	TUE	WED
1	2	3	4
8	9	10	11
15	16	17	18
22	23	24	25 *Christmas Day*
29	30	31	1
5	6	7	8

December 2019

THU	FRI	SAT
5	6	7
12	13	14
19	20	21
26	27	28
2	3	4
9	10	11

November 2019

SUN	MON	TUE	WED	THU	FRI	SAT
					1	2
3	4	5	6	7	8	9
10	11	12	13	14	15	16
17	18	19	20	21	22	23
24	25	26	27	28	29	30

notes

_____

_____

_____

_____

_____

_____

_____

_____

_____

_____

_____

_____

January 2020

SUN	MON	TUE	WED	THU	FRI	SAT
			1	2	3	4
5	6	7	8	9	10	11
12	13	14	15	16	17	18
19	20	21	22	23	24	25
26	27	28	29	30	31	

WEEK OF :………………..

SUBJECT……………………….	MONDAY Date…………………………	TUESDAY Date……………………………

WEDNESDAY Date...........	THURSDAY Date...............	FRIDAY Date...............

WEEK OF :………………..

SUBJECT………………………	MONDAY Date…………………………	TUESDAY Date…………………………

WEDNESDAY Date...........	THURSDAY Date...............	FRIDAY Date...............

WEEK OF :………………..

SUBJECT………………………	MONDAY Date…………………………	TUESDAY Date…………………………

WEDNESDAY Date...........	THURSDAY Date...............	FRIDAY Date...............

WEEK OF :..................

SUBJECT..........................	MONDAY Date............................	TUESDAY Date............................

WEDNESDAY Date............	THURSDAY Date................	FRIDAY Date................

WEEK OF :...................

SUBJECT...........................	MONDAY Date...........................	TUESDAY Date...........................

WEDNESDAY Date............	THURSDAY Date...............	FRIDAY Date...............

January 2020

SUN	MON	TUE	WED
29	30	31	1 *New Year's Day*
5	6	7	8
12	13	14	15
19	20 *Martin Luther King J Day*	21	22
26	27	28	29
2	3	4	5

January 2020

THU	FRI	SAT
2	3	4
9	10	11
16	17	18
23	24	25
30	31	1
6	7	8

December 2019

SUN	MON	TUE	WED	THU	FRI	SAT
1	2	3	4	5	6	7
8	9	10	11	12	13	14
15	16	17	18	19	20	21
22	23	24	25	26	27	28
29	30	31				

notes

_____

_____

_____

_____

_____

_____

_____

_____

_____

_____

_____

_____

February 2020

SUN	MON	TUE	WED	THU	FRI	SAT
						1
2	3	4	5	6	7	8
9	10	11	12	13	14	15
16	17	18	19	20	21	22
23	24	25	26	27	28	29

WEEK OF :....................

SUBJECT............................	MONDAY Date..............................	TUESDAY Date..............................

WEDNESDAY Date..........	THURSDAY Date...............	FRIDAY Date...............
_____	_____	_____
_____	_____	_____
_____	_____	_____
_____	_____	_____
_____	_____	_____
_____	_____	_____
_____	_____	_____
_____	_____	_____
_____	_____	_____
_____	_____	_____
_____	_____	_____
_____	_____	_____
_____	_____	_____
_____	_____	_____
_____	_____	_____
_____	_____	_____
_____	_____	_____
_____	_____	_____
_____	_____	_____
_____	_____	_____
_____	_____	_____
_____	_____	_____
_____	_____	_____
_____	_____	_____
_____	_____	_____
_____	_____	_____
_____	_____	_____

WEEK OF :.................

SUBJECT........................	MONDAY Date...........................	TUESDAY Date...........................

WEDNESDAY Date............	THURSDAY Date...............	FRIDAY Date...............

WEEK OF :.................

SUBJECT.........................	MONDAY Date.........................	TUESDAY Date.........................

WEDNESDAY Date............	THURSDAY Date...............	FRIDAY Date...............
_____	_____	_____
_____	_____	_____
_____	_____	_____
_____	_____	_____
_____	_____	_____
_____	_____	_____
_____	_____	_____
_____	_____	_____
_____	_____	_____
_____	_____	_____
_____	_____	_____
_____	_____	_____
_____	_____	_____
_____	_____	_____
_____	_____	_____
_____	_____	_____
_____	_____	_____
_____	_____	_____
_____	_____	_____
_____	_____	_____
_____	_____	_____
_____	_____	_____
_____	_____	_____
_____	_____	_____
_____	_____	_____
_____	_____	_____
_____	_____	_____
_____	_____	_____

WEEK OF :..................

SUBJECT..........................	MONDAY Date............................	TUESDAY Date.............................

WEDNESDAY Date...........	THURSDAY Date...............	FRIDAY Date...............
_____	_____	_____
_____	_____	_____
_____	_____	_____
_____	_____	_____
_____	_____	_____
_____	_____	_____
_____	_____	_____
_____	_____	_____
_____	_____	_____
_____	_____	_____
_____	_____	_____
_____	_____	_____
_____	_____	_____
_____	_____	_____
_____	_____	_____
_____	_____	_____
_____	_____	_____
_____	_____	_____
_____	_____	_____
_____	_____	_____
_____	_____	_____
_____	_____	_____
_____	_____	_____
_____	_____	_____
_____	_____	_____
_____	_____	_____

February 2020

SUN	MON	TUE	WED
26	27	28	29
2	3	4	5
9	10	11	12
16	17 *Presidents' Day*	18	19
23	24	25	26
1	2	3	4

February 2020

THU	FRI	SAT
30	31	1
6	7	8
13	14 *Valentine's Day*	15
20	21	22
2/	28	29
5	6	7

January 2020

SUN	MON	TUE	WED	THU	FRI	SAT
			1	2	3	4
5	6	7	8	9	10	11
12	13	14	15	16	17	18
19	20	21	22	23	24	25
26	27	28	29	30	31	

notes

_____
_____
_____
_____
_____
_____
_____
_____
_____
_____
_____
_____
_____

March 2020

SUN	MON	TUE	WED	THU	FRI	SAT
1	2	3	4	5	6	7
8	9	10	11	12	13	14
15	16	17	18	19	20	21
22	23	24	25	26	27	28
29	30	31				

WEEK OF :..................

SUBJECT...........................	MONDAY Date...........................	TUESDAY Date...........................

WEDNESDAY Date..........	THURSDAY Date...............	FRIDAY Date...............

WEEK OF :.................

SUBJECT...........................	MONDAY Date............................	TUESDAY Date............................

WEDNESDAY Date............	THURSDAY Date...............	FRIDAY Date...............
_____	_____	_____
_____	_____	_____
_____	_____	_____
_____	_____	_____
_____	_____	_____
_____	_____	_____
_____	_____	_____
_____	_____	_____
_____	_____	_____
_____	_____	_____
_____	_____	_____
_____	_____	_____
_____	_____	_____
_____	_____	_____
_____	_____	_____
_____	_____	_____
_____	_____	_____
_____	_____	_____
_____	_____	_____
_____	_____	_____
_____	_____	_____
_____	_____	_____
_____	_____	_____
_____	_____	_____
_____	_____	_____

WEEK OF :..................

SUBJECT.........................	MONDAY Date...........................	TUESDAY Date...........................

WEDNESDAY Date............	THURSDAY Date...............	FRIDAY Date...............
_____	_____	_____
_____	_____	_____
_____	_____	_____
_____	_____	_____
_____	_____	_____
_____	_____	_____
_____	_____	_____
_____	_____	_____
_____	_____	_____
_____	_____	_____
_____	_____	_____
_____	_____	_____
_____	_____	_____
_____	_____	_____
_____	_____	_____
_____	_____	_____
_____	_____	_____
_____	_____	_____
_____	_____	_____
_____	_____	_____
_____	_____	_____
_____	_____	_____
_____	_____	_____
_____	_____	_____
_____	_____	_____
_____	_____	_____

WEEK OF :.................

SUBJECT...........................	MONDAY Date..........................	TUESDAY Date..........................

WEDNESDAY Date...........	THURSDAY Date...............	FRIDAY Date...............

March 2020

SUN	MON	TUE	WED
1	2	3	4
8	9	10	11
15	16	17 *St. Patrick's Day*	18
22	23	24	25
29	30	31	1
5	6	7	8

March 2020

THU	FRI	SAT
5	6	7
12	13	14
19	20	21
26	27	28
2	3	4
9	10	11

February 2020

SUN	MON	TUE	WED	THU	FRI	SAT
						1
2	3	4	5	6	7	8
9	10	11	12	13	14	15
16	17	18	19	20	21	22
23	24	25	26	27	28	29

notes

_____
_____
_____
_____
_____
_____
_____
_____
_____
_____
_____
_____
_____

April 2020

SUN	MON	TUE	WED	THU	FRI	SAT
			1	2	3	4
5	6	7	8	9	10	11
12	13	14	15	16	17	18
19	20	21	22	23	24	25
26	27	28	29	30		

WEEK OF :...................

SUBJECT...........................	MONDAY Date...........................	TUESDAY Date...........................

WEDNESDAY Date............	THURSDAY Date...............	FRIDAY Date...............

WEEK OF :..................

SUBJECT..........................	MONDAY Date.............................	TUESDAY Date.............................

WEDNESDAY Date...........	THURSDAY Date...............	FRIDAY Date...............

WEEK OF :....................

SUBJECT..........................	MONDAY Date..........................	TUESDAY Date..........................

WEDNESDAY Date............	THURSDAY Date...............	FRIDAY Date...............
_____	_____	_____
_____	_____	_____
_____	_____	_____
_____	_____	_____
_____	_____	_____
_____	_____	_____
_____	_____	_____
_____	_____	_____
_____	_____	_____
_____	_____	_____
_____	_____	_____
_____	_____	_____
_____	_____	_____
_____	_____	_____
_____	_____	_____
_____	_____	_____
_____	_____	_____
_____	_____	_____
_____	_____	_____
_____	_____	_____
_____	_____	_____
_____	_____	_____
_____	_____	_____
_____	_____	_____
_____	_____	_____
_____	_____	_____

WEEK OF :..................

SUBJECT...........................	MONDAY Date............................	TUESDAY Date............................

WEDNESDAY Date...........	THURSDAY Date...............	FRIDAY Date...............

WEEK OF :....................

SUBJECT............................	MONDAY Date............................	TUESDAY Date............................

WEDNESDAY Date............	THURSDAY Date...............	FRIDAY Date...............

April 2020

SUN	MON	TUE	WED
29	30	31	1
5	6	7	8
12 *Easter*	13	14	15
19	20	21	22
26	27	28	29
3	4	5	6

April 2020

THU	FRI	SAT
2	3	4
9	10	11
16	17	18
23	24	25
30	1	2
7	8	9

March 2020

SUN	MON	TUE	WED	THU	FRI	SAT
1	2	3	4	5	6	7
8	9	10	11	12	13	14
15	16	17	18	19	20	21
22	23	24	25	26	27	28
29	30	31				

notes

_____

_____

_____

_____

_____

_____

_____

_____

_____

_____

_____

_____

May 2020

SUN	MON	TUE	WED	THU	FRI	SAT
					1	2
3	4	5	6	7	8	9
10	11	12	13	14	15	16
17	18	19	20	21	22	23
24	25	26	27	28	29	30
31						

WEEK OF :...................

SUBJECT..........................	MONDAY Date...........................	TUESDAY Date............................

WEDNESDAY Date............	THURSDAY Date...............	FRIDAY Date...............

WEEK OF :..................

SUBJECT..........................	MONDAY Date..........................	TUESDAY Date..........................

WEDNESDAY Date............	THURSDAY Date...............	FRIDAY Date...............

WEEK OF :...................

SUBJECT...........................	MONDAY Date.............................	TUESDAY Date.............................

WEDNESDAY Date............	THURSDAY Date...............	FRIDAY Date...............

WEEK OF :..................

SUBJECT...........................	MONDAY Date............................	TUESDAY Date............................

WEDNESDAY Date..........	THURSDAY Date..............	FRIDAY Date..............

May 2020

SUN	MON	TUE	WED
26	27	28	29
3	4	5	6
10 *Mother's Day*	11	12	13
17	18	19	20
24	25 *Memorial Day*	26	27
31	1	2	3

May 2020

THU	FRI	SAT
30	1	2
7	8	9
14	15	16
21	22	23
28	29	30
4	5	6

April 2020

SUN	MON	TUE	WED	THU	FRI	SAT
			1	2	3	4
5	6	7	8	9	10	11
12	13	14	15	16	17	18
19	20	21	22	23	24	25
26	27	28	29	30		

notes

_____

_____

_____

_____

_____

_____

_____

_____

_____

_____

_____

June 2020

SUN	MON	TUE	WED	THU	FRI	SAT
	1	2	3	4	5	6
7	8	9	10	11	12	13
14	15	16	17	18	19	20
21	22	23	24	25	26	27
28	29	30				

WEEK OF :..................

SUBJECT..........................	MONDAY Date..............................	TUESDAY Date..............................

WEDNESDAY Date…………	THURSDAY Date……………	FRIDAY Date……………

WEEK OF :..................

SUBJECT...........................	MONDAY Date............................	TUESDAY Date............................

WEDNESDAY Date...........	THURSDAY Date..............	FRIDAY Date..............

WEEK OF :...................

SUBJECT...........................	MONDAY Date...........................	TUESDAY Date...........................

WEDNESDAY Date............	THURSDAY Date................	FRIDAY Date................

WEEK OF :..................

SUBJECT...........................	MONDAY Date...........................	TUESDAY Date...........................

WEDNESDAY Date...........	THURSDAY Date...............	FRIDAY Date...............

WEEK OF :.................

SUBJECT.........................	MONDAY Date...........................	TUESDAY Date...........................

WEDNESDAY Date...........	THURSDAY Date...............	FRIDAY Date...............

June 2020

SUN	MON	TUE	WED
31	1	2	3
7	8	9	10
14	15	16	17
21 *Father's Day*	22	23	24
28	29	30	1
5	6	7	8

June 2020

THU	FRI	SAT
4	5	6
11	12	13
18	19	20
25	26	27
2	3	4
9	10	11

May 2020

SUN	MON	TUE	WED	THU	FRI	SAT
					1	2
3	4	5	6	7	8	9
10	11	12	13	14	15	16
17	18	19	20	21	22	23
24	25	26	27	28	29	30
31						

notes

_____

_____

_____

_____

_____

_____

_____

_____

_____

_____

_____

July 2020

SUN	MON	TUE	WED	THU	FRI	SAT
			1	2	3	4
5	6	7	8	9	10	11
12	13	14	15	16	17	18
19	20	21	22	23	24	25
26	27	28	29	30	31	

WEEK OF :.................

SUBJECT...........................	MONDAY Date...........................	TUESDAY Date...........................

WEDNESDAY Date..........	THURSDAY Date..............	FRIDAY Date..............

WEEK OF :..................

SUBJECT...........................	MONDAY Date.............................	TUESDAY Date.............................

WEDNESDAY Date............	THURSDAY Date...............	FRIDAY Date...............

WEEK OF :.................

SUBJECT...........................	MONDAY Date...........................	TUESDAY Date...........................

WEDNESDAY Date............	THURSDAY Date...............	FRIDAY Date................

WEEK OF :.................

SUBJECT...........................	MONDAY Date............................	TUESDAY Date.............................

WEDNESDAY Date…………	THURSDAY Date……………	FRIDAY Date……………

_____

_____

_____

_____

_____

_____

_____

_____

_____

_____

_____

_____

_____

_____

_____

_____

_____

_____

_____

_____

_____

_____

_____

_____

_____

_____

_____

_____

July 2020

SUN	MON	TUE	WED
28	29	30	1
5	6	7	8
12	13	14	15
19	20	21	22
26	27	28	29
2	3	4	5

July 2020

THU	FRI	SAT
2	3	4 *Independence Day*
9	10	11
16	17	18
23	24	25
30	31	1
6	7	8

June 2020

SUN	MON	TUE	WED	THU	FRI	SAT
	1	2	3	4	5	6
7	8	9	10	11	12	13
14	15	16	17	18	19	20
21	22	23	24	25	26	27
28	29	30				

notes

_____
_____
_____
_____
_____
_____
_____
_____
_____
_____
_____
_____

August 2020

SUN	MON	TUE	WED	THU	FRI	SAT
						1
2	3	4	5	6	7	8
9	10	11	12	13	14	15
16	17	18	19	20	21	22
23	24	25	26	27	28	29
30	31					

WEEK OF :..................

SUBJECT...........................	MONDAY Date.............................	TUESDAY Date.............................

WEDNESDAY Date............	THURSDAY Date...............	FRIDAY Date................

WEEK OF :..................

SUBJECT...........................	MONDAY Date.............................	TUESDAY Date.............................

WEDNESDAY Date............	THURSDAY Date...............	FRIDAY Date...............

WEEK OF :..................

SUBJECT..........................	MONDAY Date............................	TUESDAY Date............................

WEDNESDAY Date............	THURSDAY Date...............	FRIDAY Date...............

WEEK OF :..................

SUBJECT...........................	MONDAY Date............................	TUESDAY Date...........................
	_____	_____
	_____	_____
	_____	_____
	_____	_____
	_____	_____
	_____	_____
	_____	_____
	_____	_____
	_____	_____
	_____	_____
	_____	_____
	_____	_____
	_____	_____
	_____	_____
	_____	_____
	_____	_____
	_____	_____
	_____	_____
	_____	_____
	_____	_____
	_____	_____

WEDNESDAY Date............	THURSDAY Date................	FRIDAY Date................

CONTACT *List*

Names	Mobile	Home Phone	E-mail

Made in the USA
Las Vegas, NV
27 August 2021

29097313R00096